ARMOR UP

Exoskeletons, Scales, and Shells

Enslow Publishing
101 W. 23rd Street
Suite 240
New York, NY 10011
USA
enslow.com

Emma Carlson Berne
and Susan K. Mitchell

These books are dedicated to Emily, who inspired the author.
—Susan K. Mitchell

Published in 2020 by Enslow Publishing, LLC.
101 W. 23rd Street, Suite 240, New York, NY 10011

Library of Congress Cataloging-in-Publication Data

Names: Berne, Emma Carlson, author. | Mitchell, Susan K., author.
Title: Armor up : exoskeletons, scales, and shells / Emma Carlson Berne and Susan K. Mitchell.
Description: New York : Enslow Publishing, 2020. | Series: Animal defense! | Audience: Grade 3-6. | Includes bibliographical references and index.
Identifiers: LCCN 2018046709| ISBN 9781978507142 (library bound) | ISBN 9781978508071 (paperback)
Subjects: LCSH: Armored animals—Juvenile literature. | Animal defenses—Juvenile literature.
Classification: LCC QL940 .B47 2020 | DDC 591.47/7—dc23
LC record available at https://lccn.loc.gov/2018046709

Printed in the United States of America

To Our Readers: We have done our best to make sure all website addresses in this book were active and appropriate when we went to press. However, the author and the publisher have no control over and assume no liability for the material available on those websites or on any websites they may link to. Any comments or suggestions can be sent by email to customerservice@enslow.com.

Portions of this book originally appeared in *Animals with Awesome Armor: Shells, Scales, and Exoskeletons.*

Photo Credits: Cover, p. 1 (sea turtle in water) Kjeld Friis/Shutterstock.com; cover, p. 1 (reptile scales) G-ZStudio/Shutterstock.com; p. 6 Sista Vongjintanaruks/Shutterstock .com; p. 9 Fotogrin/Shutterstock.com; p. 11 NokHoOkNoi/Shutterstock.com; p. 13 Ksena Shu/Shutterstock.com; p. 15 Tim Zurowski/Shutterstock.com; p. 19 © iStockphoto. com/MirianeDL; p. 21 Eugene Troskie/Shutterstock.com; p. 23 Barry Blackburn/ Shutterstock.com; p. 26 Magdalena Paluchowska/Shutterstock.com; p. 28 Kevin H Knuth/Shutterstock.com; p. 31 Jeff Rotman/Alamy Stock Photo; p. 33 Ezume Images/ Shutterstock.com; p. 37 George Grall/National Geographic Image Collection/Getty Images; p. 40 Renata Sedmakova/Shutterstock.com; p. 42 Fotokostic/Shutterstock.com.

Contents

Introduction

When people think of the word "armor," they probably picture a knight from the Middle Ages wearing chain mail and a big metal helmet. Or maybe they imagine a modern soldier wearing a bulletproof vest and a helmet covered in **camouflage**.

One thing they probably don't imagine is an armadillo. They probably aren't thinking of a clam or even a snail. But these animals all wear armor, just like knights did and soldiers still do. Instead of putting on their armor before a battle, these animals wear their armor all the time. It's a part of their bodies. Shells, scales, bony plates, and tough skin help reptiles; **mollusks**, such as snails and clams; and even some mammals survive. Insects and crustaceans, such as lobsters, have **exoskeletons**. Their entire bodies are armor! The exoskeleton is like wearing bones on the outside of the body. A hard covering surrounds the soft inside of the body.

Just like knights going into battle, animals use armor to protect themselves from harm, especially from predators. Armadillos have heavy bands of armor across their backs and sides. When a bird of prey swoops down to try to grab an armadillo, the bird's talons can't grab the armor bands. The armor makes the armadillo too heavy and awkward to grab. The bird might fly away. Armor for the win!

Even small animals can have unusually tough armor. The scaly-foot snail lives on the bottom of the Indian Ocean. This little snail has to defend itself against crabs. It has evolved a special three-layer shell that is hard for crabs to crack when they squeeze it with their claws. The scaly-foot snail even has armor on its foot. Scales help protect the delicate foot against other snails that are venomous.

Some animals have armor only on parts of their bodies. They might curl up like a ball or pull back into a shell when a predator comes near. They shield the soft, unprotected parts of their bodies,

The heavily armored alligator snapping turtle can be found in lakes, swamps, and large rivers in the southeastern United States.

showing only the armored portions. Some animals have spikes or growths on their armor to defend even better against predators. The alligator snapping turtle has jagged spikes on its shell and an armored tail. Of course, armor doesn't always work. Sometimes, predators can still kill animals with armor. Scientists know that the jaguar has especially heavy muscles in its head and thick teeth. Recently, scientists suggested that these **adaptations** help the jaguar bite through armor to kill prey such as turtles, armadillos, and caimans, which are relatives of the alligator.

Armor helps protect animals against more than just predators. Shells, scales, and tough skin shield animals from extreme heat, provide shade, and even protect against insect stings or bites. Armored animals have an extra layer of protection in an often fierce, dangerous world. They can live another day to eat instead of being eaten.

Chapter 1

Armor of All Sorts

All animal armor is tough, but it is not all the same. Some armor is bone. Some is a substance called **keratin**. Insect shells are often composed of a material called **chitin**. Often snail shells are made of **calcium carbonate**. Some of these materials are very thick and rigid. Some are flexible. Some animals grow several coverings of armor in their lifetimes. Others have permanent armor.

Bone is one of the toughest types of armor. Bones are strong enough to make up the skeleton inside humans and many animals. They can also be a tough covering on the outside. Bony armor is usually connected to the outside of an animal's skin. That means that the animal cannot crawl out of its armor. Turtles are one type of animal with this kind of armor.

The great thing about bony armor is that it lasts forever. An animal with armor made of bone never has to grow new armor. Bone armor grows with the animal. But even tough bony armor has weaknesses. For example, bone has nerve endings. This means that

Tortoises are relatives of turtles that live only on land. Turtles live partly in water and partly on land. Tortoises, like turtles, have bony armor.

an animal with bony armor can feel pressure and pain if the armor is damaged.

Not Chain Mail: Armor Materials

Bony armor is often covered in keratin. That is the same type of material that makes up human fingernails. Keratin by itself is also a kind of armor. It grows like bony armor, but it is much softer than bone. It bends easily. Many reptiles, such as snakes, have tough scales made of keratin.

Many insects and crabs have exoskeletons made of chitin. It is very strong, but unlike bone, it does

Fun Fact!

Vets use drills, screws, wire, and glue to hold together broken turtle shells until they heal.

A cockroach molts, shedding its old exoskeleton to reveal a new one underneath.

not grow. Animals with chitin armor have to keep growing new armor. As the animal grows, its old armor splits and the animal sheds it. This is called molting. Underneath is a new suit of chitin armor ready to protect the animal.

Some animal armor also contains calcium carbonate. This is a mineral found in shelled sea animals and snails. An animal with this type of armor grows its own shell. Its body makes the calcium carbonate, which forms a shell around the animal. This makes the armor harder, stiffer, and stronger than armor made only from chitin. Animals with this type of armor do not have to molt. They just keep adding calcium to their shell.

Dinosaur Armor

Some of the most well-known rough and tough armored animals were the dinosaurs. But just because they were big did not mean they were always safe. Plant-eating dinosaurs were especially at risk. They could be huge but

Ancient Armor

In 2011, a miner in Canada accidentally dug up what later turned out to be dinosaur fossils. He had found the fossilized remains of half of the whole body, including armor, skin, and tendons. The dinosaur was a heavily armored type known as a nodosaur. The nodosaur was covered with thorny armored plates from head to tail and had a pair of horns jutting out of its shoulders.

These two ankylosaur models from a theme park showcase the dinosaurs' impressive armor.

usually did not have sharp claws or teeth for defense. Instead, they had hard, bony plates or super-thick skin for protection.

Many of today's armored animals can trace their history back even before the age of the dinosaurs. And their armor has not changed much. The same kind of armor that protected them millions of years ago still works well today.

Chapter 2

Banding Together

Not many armored animals are mammals. Reptiles are often armored, and so are insects and crustaceans. But mammals often have different ways to defend themselves. Not the armadillo. The armadillo is a mammal *and* it is covered in tough armor. The word "armadillo" means "little armored one" in Spanish. There are more than twenty types of armadillos. Some, such as the giant armadillo, are the size of a large dog. Others, such as the pink fairy armadillo, are tiny—only about 6 inches (15 centimeters) long.

The armadillo's armor is called a **carapace**. It is not a solid shell like a turtle's. An armadillo's armor is made up of sections called bands. Armadillos are grouped by

An armadillo tests the water in the Rio Grande Valley of Texas. Armadillos can live in humid or dry warm climates.

how many bands of armor they have—usually three, six, or nine. The bands are made up of bone. The bony bands overlap each other. They are covered with hard scales made of keratin.

The armadillo does have some hair. The hair mostly grows on the armadillo's sides and underside. Sometimes this hair is called "curb-feelers." Like insect feelers, it helps the armadillo "feel" objects in the dark.

The most common armadillo is the nine-banded armadillo. It can be found all over South America. It is also very easy to find in the southern United States, especially in Texas and Florida. Since armor does not

Fun Fact!

Nine-banded armadillos can jump as much as 3 or 4 feet (about 1 meter) straight up into the air when they are scared.

help keep an armadillo's body very warm, they live in warm areas.

Made for Digging

Armadillos are born diggers. From the moment they are born, they have the need to dig. An armadillo's body is perfect for digging. It is low to the ground. Its legs are very short. On the end of an armadillo's front legs are huge, strong claws.

These claws are good for digging for food. Armadillos eat mostly insects. They also sometimes eat roots or bird eggs. Armadillos have even been known to nibble on dead animals. They have very small eyes and do not see well. They use their flat snout and claws to push along the ground looking for a bite to eat.

Don't Eat Armadillos!

Armadillos can make people sick. They often carry the bacteria that causes Hansen's disease, also known as **leprosy**. Leprosy is a bacterial infection that can affect people's nerves, skin, and eyes. Most of the time, armadillos with leprosy do not make people sick. But recently, scientists found that when people kill and eat armadillos, they can contract the disease. In Brazil, armadillo is a popular animal to eat. People often enjoy a dish made of raw armadillo liver and onions. Researchers warn that Brazilians who eat this dish and other food made from armadillo meat are at higher risk of getting leprosy.

While an armadillo's sight and hearing may be bad, its sense of smell is super. An armadillo can smell a tasty treat that is several inches under the ground. When an armadillo finds food, it starts digging. It uses a long, sticky tongue to grab the food. The armadillo's tongue does most of the work when eating. It grabs insects or bites of other food. It also mashes up the food before the armadillo swallows. Anything that cannot be mashed by the tongue gets ground up by the armadillo's teeth.

Armadillos also dig to make tunnel-like homes under the earth, called burrows. Unlike some burrowing animals that may only have one or two burrows, an armadillo might have many. Some burrows are used as a living area. Others are used to escape danger. Armadillos also dig burrows to keep warm during cooler weather.

Ball Up

Most of the time, the armadillo's armor keeps it safe. Unfortunately, it is not a perfect defense. Armadillo babies, called pups, are born with soft armored bands that feel like leather. The armor hardens as the pup

A three-banded armadillo rolls into a ball to protect its soft underbelly and organs. The head and tail are the two long parts at the top.

grows older. But even then, the armor covers only the armadillo's head, shoulders, back, and tail. Its belly, however, is not protected at all.

The first thing an armadillo does when it feels threatened is hide or run away. It may try to dig into the ground very quickly. If there is no time to dig, it will lie close to the ground instead. This is a way to try to protect its soft belly and legs.

The three-banded armadillo can protect its belly in a very interesting way. Its bands of armor are very flexible. They allow the armadillo to curl up into a tight ball. It pulls its legs close to its body. It tucks its head and tail inside also.

The pink fairy armadillo has another strange way of protecting itself. This tiny armadillo has a hard plate covering its rear end. When in danger, the pink fairy armadillo dives headfirst into a burrow. Its rear plate acts like a shield.

The nine-banded armadillo is one of the few armadillos that can swim away from danger. It can fill its stomach with air and float away. It can also hold its breath and walk along the bottom of a stream or river. The nine-banded armadillo can hold its breath for a very long time.

A Sharp Critter

Armadillos are not the only mammal with armor. The pangolin (PANG-ga-lin) of Africa has even tougher scaly armor. This animal is also called a scaly anteater. These strange creatures live very much like armadillos. Pangolins dig to find food. They also live in burrows.

The armor of the pangolin is also a lot like an armadillo's. It is made of many hard overlapping scales. The scales are made of keratin just like an armadillo's.

The pangolin is a unique looking animal with sharp scales. It can also release a stinky liquid to ward off attackers.

There is one huge difference between the scales of a pangolin and an armadillo, however. The pangolin's scales are razor sharp! When frightened, it rolls itself into a tight ball. The super-sharp edges of these scales would cut any predator that tried to pry open the pangolin. This animal's armor can be a weapon as well as a defense.

Chapter 3

Swell Shells

A turtle shell is a truly amazing piece of armor. It protects the turtle during its entire life. It can heal if it gets broken. And turtle shells have been tested by time. Turtles belong to the oldest reptile group. They have been on Earth since the time of the dinosaurs—more than two hundred million years ago. Turtles have evolved very little since their earliest days on Earth.

Turtles go by many names. This usually depends on where they live. "Turtle" usually refers to the animals that live in water. They have flipper-like legs and feet that help them swim. "Tortoise" is the name used for turtles that usually live only on land. Their shells are more dome-shaped than the flatter turtle shells. Tortoises also have rounder, thicker legs and feet than turtles. Somewhere in between are the terrapins. These animals can live equally in water and on land.

Complete Protection

A turtle's shell is made up of two parts. The top part is the carapace. The bottom part is called the plastron. Unlike an armadillo, a turtle's belly is also protected. The entire shell is attached to the turtle's backbone and ribs.

The shell is made of very hard bone and is covered in plates of keratin called scutes. The shell is not one large bone, however. It is actually made up of fifty to sixty bones that are joined together. Just like the bones of any animal, the shell grows with the turtle.

A female Eastern painted turtle hides in her shell to protect herself.

Only the turtle's head, legs, and tail are not covered by its shell. To protect these areas, most types of turtles can pull their legs and heads inside their shells. Other turtles can only pull their long necks to the side. Then their heads fit neatly under the front of the shell. Box turtles have a special muscle in their plastron. They can use this muscle like a hinge to close the top and bottom shells

There's No Place Like Home

Box turtles spend their entire lives within about 250 yards (229 meters) of the place where they were born. They know where to find food in their territory with a mental map they make for themselves. If box turtles are moved from their home territory, they will try to walk back. For this reason, wildlife experts tell people who find box turtles crossing roads to not move them far away. Just help them across the road. The turtle will help itself after that.

Fun Fact!

The oldest turtle known to humans, named Tu'i Malila, was 188 years old.

tightly together. Sea turtles are the only turtles that cannot pull themselves into their shells.

A turtle's shell is great protection from predators. It also helps a turtle in other ways. A sea turtle's smooth, flat shell helps it dive and swim. A large, domed shell like that of the Galapagos tortoise helps shade the turtle from the hot sun. Many shells also blend in with the turtle's environment to help it hide.

A Dangerous Time

Since turtles are reptiles, they lay eggs. After female turtles lay their eggs, most of them leave the nest.

A baby sea turtle crawls across the sand to the ocean soon after it hatches.

This means that baby turtles are on their own from the moment they hatch. Even with their small shells, they are almost defenseless. Many baby turtles are eaten by predators before they have a chance to leave the nest.

For the lucky babies who survive, growing up to adulthood can take a very long time. Some turtles live to be more than a hundred years old. As they grow, the turtles' armor keeps growing and protecting them.

Chapter 4

A Really Cool Crab

A spiky shell trundles along the ocean floor, navigating with its many legs. It might look like a walking helmet. Or it might look like a strange relative of a crab or lobster. In fact, this armored creature is called a horseshoe crab. But horseshoe crabs are actually related to scorpions and spiders. And horseshoe crabs are even older than the dinosaurs—they have been on Earth for about 450 million years!

The tough shell of a horseshoe crab is made of chitin. This is the same type of material that makes up the hard outer covering of many insects. Unlike keratin or bone, chitin does not grow. The horseshoe crab molts several times during its first years of life.

Empty horseshoe crab shells often wash up on beaches after the horseshoe crab has molted.

When a horseshoe crab molts, the old shell splits open and the horseshoe crab crawls out of it. Until the new shell underneath hardens, the horseshoe crab is very soft. It has no armor and is in danger. During this time, horseshoe crabs hide in the mud. Each time a horseshoe crab grows a new set of armor, it is bigger than the last. It takes from seven to ten years for a horseshoe crab to reach its full size. Then it stops molting.

Three Body Parts

A horseshoe crab's body has three parts. The large part in the front that includes the head is called the **cephalothorax**. Behind that is the **abdomen**. The last part is a very long, sharp tail called a telson. A horseshoe crab's armor has two different parts—one that covers the front area and another that covers the back area.

The two pieces of armor are hooked together by a hinge. This lets the horseshoe crab bend in the middle. Being able to bend helps it dig. It also helps the horseshoe crab do a flip. That is handy if it gets washed onto its back by a wave. The horseshoe crab bends in the middle and uses its tail to flip itself armor-side up.

Valuable Blood

One of the most unique things about horseshoe crabs is their blood—it's blue! Unlike human blood, a horseshoe

Fun Fact!

The horseshoe crab's tail looks like a stinger, but it's not. This animal uses its tail to flip itself over if it gets overturned.

crab's blood has copper in it. The copper causes the blood to turn blue if it is exposed to air.

The blood of a horseshoe crab can also protect it against germs. If the horseshoe crab gets wounded, the blood forms a hard plug called a clot. The clot keeps the horseshoe crab from getting a serious infection.

Scientists use horseshoe crab blood to test for dangerous germs and to test some medicines and medical equipment for safety. If the blood forms a clot on the material being tested, scientists know the material

A Tough Nut

Giant coconut crabs live on the islands of the Pacific Ocean. This 3-foot-long (1-meter) crab climbs coconut palm trees to clip the tough fruit and crack into it with its long pincers. The crab's brown shell even looks like a coconut. It provides camouflage when the crab is high in the tree. When the babies hatch, their shells are soft. So the baby crab pulls a small coconut shell onto its back. It wears the shell until its own armor is hard and thick enough.

Blood taken from horseshoe crabs is used for medical research. The horseshoe crabs are not harmed. They will be released once scientists have collected the blood they need.

is not safe for humans. Horseshoe crabs are not hurt or killed for their blood. Only a small amount of blood is taken. Then they are released back into the ocean.

The chitin that forms the armor is also used by scientists. They use it to make special threads called sutures, which are used by doctors to sew up cuts and wounds. Cuts sewn up by the special chitin thread heal much faster than normal. Horseshoe crab chitin is also used to make bandages for burn victims.

Chapter 5

Tough Armor, Lots of Legs

Living under a rock in some old dead leaves is a bug-like creature with many legs. It is a millipede. Millipedes are part of a group of animals called diplopods. The word means "double legged." Millipedes are protected by their exoskeleton, the hard outer skin that covers the animal's whole body.

Millipedes have many body parts. These parts are called **segments**. Each segment has two pairs of legs attached to it. This is where the name "diplopod" comes in. The number of segments depends on the size of the millipede. The smallest millipede is only about an inch (2½ cm)

Millipedes come in many colors. This millipede's exoskeleton is orange, black, and yellow. They can be brown, ivory, or red as well.

long. These small millipedes usually have around six or seven body segments. The largest millipede in the world is the giant African millipede. It can grow to be more than a foot (30½ cm) long! This huge millipede can have several hundred legs. That means a lot of segments!

The name "millipede" means "thousand foot." The millipede has many legs, but nowhere near a thousand.

Fun Fact!

Millipedes are known as "recyclers." They feed on damp, rotting leaves that have fallen to the ground. They keep dead leaves from piling up on the forest floor.

Most millipedes have only several hundred. The centipede ("hundred foot") is a close cousin of the millipede.

Despite having so many legs, the millipede is actually a slow-moving animal. So it needs its tough armor to protect itself from predators that would love to eat it.

Armor Defense and More

Baby millipedes hatch after several weeks inside an egg. Then they go through a second life stage, becoming nymphs.

They will keep growing during this stage. As a millipede grows, it molts. Each time it molts, the millipede grows more segments and legs. Most millipedes will molt many times before they become adults.

Many predators search for millipedes that are molting. A millipede is very soft after it sheds its hard outer skin. It must wait for the new armor to harden before it is safe again. But even after a millipede's armor hardens, predators love to catch them. Like an armadillo, only the top part of a millipede is protected by armor. Its belly is soft and unprotected.

Fortunately, millipedes have more protection than just their hard armor. The millipede can roll up into a tight ball to help keep its soft belly safe. But some predators have even learned how to get around this. A mongoose, for example, smashes the millipede against a rock to break its armor.

Many millipedes also have chemical defenses. They can ooze a stinky liquid. Most predators will leave the smelly millipede alone. Some millipedes can even produce poison. Some of these poisons simply sting or burn. There is even a poison that can make a small predator fall asleep! This gives the millipede time to get away.

Other millipedes' poisons are stronger and can even kill small predators. These strong poisons are not enough to harm large predators such as birds. They are also usually not very harmful to humans. Against most predators, however, a millipede's super-tough armor is enough to keep it safe.

That's Some Backpack!

The *Acanthaspis petax* belongs to a group of insects known as assassin bugs. It hunts a particular species of ant, and when the ants are dead, it puts them on its back. *Acanthaspis petax* holds the bodies together with a sticky substance it makes. It piles so many ants on its back that it looks like a walking ant-ball. It uses these ants as both armor against its own predators and as camouflage.

More Than Just Armor

The millipede isn't the only insect with plenty of armor protection against predators. The palmetto tortoise beetle has a dome of armor on its back. It looks like a tiny walking crash helmet. These dark-purple beetles can be found in the southern United States, in places such as Florida and Texas. Palmetto tortoise beetles live anywhere that palmetto plants grow, since that is their favorite food.

This beetle has more than just its armor for protection. When the palmetto tortoise beetle is attacked, it sticks tight to whatever leaf it is eating at the time. It has special feet that make the beetle almost

A palmetto tortoise beetle clings to a leaf. Each of its feet is covered with about ten thousand sticky bristles.

impossible to move. These feet have thousands of tiny, sticky bristles, as well as an oily liquid on them. This helps the palmetto tortoise beetle clamp down tightly and stick to a leaf.

Once the beetle is stuck, it can take a lot of pulling and tugging from a predator. No matter how hard an animal pulls, the beetle will not budge. Most predators give up and find food somewhere else. When the coast is clear, the palmetto tortoise beetle pops back up and keeps on eating.

Chapter 6

Animal Armor, Human Armor

Ancient humans learned much about armor from animals. During battle, soldiers even used tough animal skins, skulls, and bones for protection. Then humans discovered an incredibly useful material—metal. They found that it made great armor. It also made better weapons than bone, rock, or wood.

Over the years, armor has changed as weapons have changed. Armor had to be built tougher as weapons became deadlier. Fighters wearing animal skin or bones would be no match for metal weapons such as swords.

The same is true today. Weapons are still changing. They have become far more deadly. Metal swords and

Metal suits of armor were effective, but they were heavy and hot for the knights wearing them.

spears are gone. Guns, rockets, and bombs have taken their place. It has been important for armor to change also. Soldiers and police officers who face these deadly weapons need armor that is better than ever.

One of the big changes is the weight of the armor. If soldiers or police today had to wear clunky metal armor, they would never catch up with the bad guys. Soldiers and police have to be able to move fast. It is important to have light but tough body armor. The armor also needs to be comfortable and cool to wear.

Tougher Than a Bullet

Kevlar is an amazingly tough fiber that was invented in the late 1960s. This material and others like it are used to make bulletproof vests. Kevlar is five times stronger than steel but very lightweight. The Kevlar fibers are woven

Military Animals Wear Armor, Too

Animals of all kinds have accompanied humans into war and when they did, they needed protection as well. Historically, people have made armor for their animals out of the same materials they used for themselves. In the early days of the first millennium, elephants wearing armored body coverings and head coverings would charge enemy soldiers. Horses carrying medieval knights into battle had their own set of full chain mail and plate armor, complete with helmets. And in modern times, dogs serving alongside combat soldiers have bulletproof vests and goggles just like their human partners.

into a very tight "net" of fabric. There are many layers of this tight net in a bulletproof vest. The layers help spread the energy of a speeding bullet evenly through the vest. The more layers a vest has, the more powerful a bullet it can stop.

A person wearing a Kevlar vest is not completely free of injury when hit by a bullet, however. The force of a bullet as it hits the vest can still knock a person to the ground. The victim may suffer severe bruising or even a few broken ribs. But those are a small price to pay for surviving a gunshot.

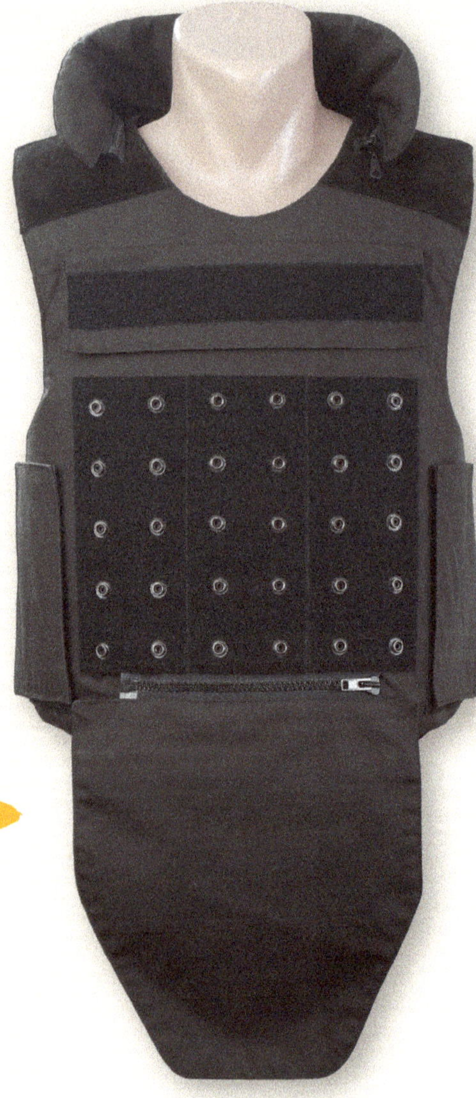

Bulletproof vests are made of Kevlar. They are much lighter than metal armor.

Kevlar is also used to make armored helmets and is used for gloves and on military vehicles.

Secrets of Animal Armor

For humans, animal armor provides more than just inspiration. Scientists often study animal armor to discover ways to improve human armor. A team of scientists from the Massachusetts Institute of Technology has been studying the structure of the shell of the scaly-foot snail. They found that the shell has three layers. When a crab squeezes the shell of the snail with its claw, the three layers help distribute the force of the impact. This makes the shell especially strong. The scientists hope that engineers can make armor, such as helmets and vests, for soldiers that has the same three-layer structure but uses man-made bulletproof materials.

Animal armor can even improve on existing human-made armor. Kevlar is a very strong and tight material, but needles can occasionally slip through, poking the wearer. Researchers discovered that a fish called the alligator gar has such tough scales that they couldn't pierce them with a steel needle. But the scales are also flexible so that the fish can move easily. Researchers are copying the structure of these strong scales using ceramic. They glue the ceramic scales onto Kevlar gloves so that not even a sharp needle can slip through.

Weak Material, Strong Shells

Researchers studying conches found that even though the shells were ten times stronger than mother-of-pearl, the proteins making up the shells were very weak and fragile. The scientists compared the proteins to chalk and Jell-O—neither of which are very good for making armor! But when they examined the proteins more closely, the scientists found that they were arranged in stacks in a way that made them extremely strong. Then the scientists were able to mimic this arrangement with man-made materials using a 3D printer. They found they could print out sheets of a strong, thin, flexible material. Scientists hope this material could be used to make sports helmets or military armor stronger and safer.

Fun Fact!

The golden tortoise beetle stacks its own poop on a forked body part that it hangs over its back. Predators don't like this poop shield and will often stay away.

Nature has had to evolve strong, special materials and structures to protect animals that might otherwise be eaten by predators. Armored animals are able to withstand many attacks that would kill other animals. Whether the armor is tough skin, bone, or stiff scales, it can give an animal the extra protection it needs to survive and can be an amazing animal defense. Humans continue to study and learn from animal armor.

Glossary

abdomen The middle section of an animal, such as an insect or crab.

adaptation A change in an animal over time that helps the animal survive.

calcium carbonate A powdery substance found in nature as chalk, limestone, or the armor of some animals, such as snails.

camouflage An animal's coloring or shape that helps it hide from predators.

carapace A hard, protective shell-like covering of an animal.

cephalothorax The front section of an animal such as a spider or crab that is made up of the head and thorax.

chitin A tough, protective material that makes up the exoskeleton of some animals.

exoskeleton The hard outer covering of some animals that provides protection.

keratin A tough material that makes up hair, fingernails, hooves, feathers, and horns.

leprosy A bacterial infection also known as Hansen's disease that can damage the nerves and affect the eyes and skin.

mollusk A member from a group of animals that includes snails and clams with a soft body usually covered by a shell.

segment One of several parts into which something is divided.

Further Reading

Books

Fortescue, Madeleine. *Super Skin: Spines, Quills, and Plates*. New York, NY: Gareth Stevens Publishing, 2016.

Kroll, Jennifer. *Showdown Animal Defenses*. Huntington Beach, CA: Teacher Created Materials, 2017.

Meister, Carl. *Armored Animals*. Minneapolis, MN: Jump! Inc., 2016.

Saidian, Siyavush. *Staying Safe with Scales and Scutes*. New York, NY: PowerKids Press, 2018.

Websites

Field Museum: Milli-Peet: Millipedes Made Easy

www.fieldmuseum.org/science/special-projects/milli-peet-class-diplopoda/milli-peet-millipedes-made-easy

Learn more about this diplopod.

National Geographic Kids: 10 Totally Awesome Facts About Turtles!

www.natgeokids.com/uk/discover/animals/sea-life/turtle-facts/

Read a general overview of this armored reptile.

San Diego Zoo Global Zoonooz: Armored Animals

zoonooz.sandiegozoo.org/2018/01/16/armored-animals/

Find out more about different animals with armor.

Index